Table of Contents

I0486992

Safety in Nuclear Power Plants!
By
Chakrapani Srinivasa

Safety in Nuclear Power Plants!

By
Chakrapani Srinivasa

Copyright 2020 Chakrapani Srinivasa

Dedicated To My Dear Parents

About the Author

Chakrapani Srinivasa (Padmaja), Freelance journalist from India possesses Bachelor degree in Engineering (B.E) and Post graduate in Business Management (MBA) with Distinction. He has worked as Associate Editor of 'Naradar' fortnightly journal in Chennai, India. He is the Senior Editor of the journal "The Divineness".

He has written many articles and e books in English through Smashwords Inc, Kindle Direct Publishing, Atlanta Publications, Cooperjal publications (UK), lulu.com, ezinearticles.com, shvoong.com, iproclaim.com (USA) and a news journal (Germany). He is the Consulting Editor: Contemporary Who's Who-Research Board of Advisers of ABI, USA.

Contributed articles, short stories and travelogues in Tamil , which are published in leading journals like Ananda Vikatan, Kumudam, Savi, Kalki, Dinamani Kadhir, Dinamani daily, Idhayam Pesukirathu, Naradar,Thisaigal, etc .Written many novels, short stories and travelogues in Tamil, which are published by pothi.com. More than 60 books written by him have been published by leading international publishers.

Preface

For economically and technologically progressing India, the nuclear storage is a challenging issue. Many aspects of Nuclear Safety practices and precautions taken in Nuclear Power Plants are discussed in this book.

Topics like Personnel's protection, Safety Drill, Safety Practices, Safety Suits, Safety Signs, Safe Transportation, Threats, Dangers, Nuclear Safety Plans and designs are dealt and analyzed interestingly.

Nuclear Storage in India

No modification or re-allocation can be done in labs, where radioactive materials are used. Radiation Safety Cell will give approval in writing for any changes in storage.

For economically and technologically progressing India, the nuclear storage is a challenging issue.

Personnel Protection:

Radiation Safety Committee is the authorized team to take care of storage and steps to avert any hazards.

Director and Radiation Safety Officer join hands with Radiation Safety Committee for all nuclear storage activities. They monitor that exposure is safe below regulatory limits for all those who are all involved in nuclear storage areas. They ensure that sufficient training in radiation safety rules and protective steps are undertaken by all users.

Radiation Protection Rules (RPR) 2004 is adhered to strictly at all places and time. They issue license to the authorized users of the radioactive materials. These authorized personnel will maintain records containing the procedures and safety rules pertaining to radioactive materials stored. They hold a degree in Science or Post graduate degree in Radiological Physics. A competent authority has to approve his qualifications. They are entrusted with responsibility like chalking out radiation protection program, which should be followed and practiced meticulously.

Safety Drill

If any radiation leak is detected or suspected, then it will be recorded for necessary action. He also takes care to see that radiation monitoring gadgets are properly fine tuned and check for its genuineness. Safety drill is conducted to all personnel to see that they manage any emergency. Communication without delay is ensured to avoid calamity. All details of doses of the workers are recorded along with the radioactive materials stored. Once in 4 months the RSO conducts review meeting to see that safety aspects are followed in storage zone. Any incidents like diarrhea, vomiting, suffocation, dizziness and reddish skin are immediately recorded and communicated to higher-ups for action.

As the total dose received by an individual is directly proportional to the time spent close to radioactive source, steps are taken by RSO to reduce that time for all engaged with radioactive materials.

Safeguard Practices

Maximization of distance from the radioactive source, shielding of Alpha radiation by air, water, paper or any material are followed strictly. Beta particles are shielded by water, glass, plastic etc. But it is difficult to escape from X-rays and Gamma rays. So, natural barriers, trenches, protective shelters are helpful to safeguard from it.

Thermo Luminescent Dosimeters, Film badges and direct reading Dosimeters are used to judge the external dose measurement on individuals. Environmental condition is monitored by Radiation Detectors. For men and women, the limit of occupational exposure is same. But if the women are conceived, then it is viewed cautiously and protected.

The Board of Radiation and Isotope Technology located in Mumbai is the authorized unit to import radioactive materials and a senior nuclear scientist possessing much experience will receive it in his name. RSO will frequently check the radioactive storage unit. The Radiation Safety Cell will be contacted by the user for disposal of radioactive materials. It will be sent back to supplier for safe disposal with correct documentation and record formalities. All details will be communicated to AERB and BRIT.

No modification or re-allocation can be done in labs, where radioactive materials are used. Radiation Safety Cell will give approval in writing for any changes in storage. Personnel Monitoring Badges or Electronic Pocket Dosimeters will be worn by Gamma Cells maintenance staff. RSO will be apprised of all works carried out by them. Those who have undergone rigorous training are allowed to operate Gamma cells.

Symbols for Safety

Radiation symbol is fixed on the rooms accommodating the Gamma cells and also on Gamma cell. The phone numbers of Safety Officers are displayed everywhere to face unexpected explosion or damages.

Safety Suits

Radiation levels in various locations are entered in register. The trained staff engaged in operation and maintenance wear full sleeves apron, mask, cap, shoe covers, gloves and plastic safety glasses.

Radiation survey is done periodically by the authorized users and audited by Radiation Safety Cell.

Disposable gloves and remote handling tongs are used for clearing the contamination. The workers are advised to change their hand gloves frequently and dispose them as radioactive waste item with a plastic bag.

Low Range Survey Meter and Liquid Scintillation Counter are used to detect contamination.

Safety Instructions

Centralized Radioactive source storage facility is available to store the disposed radioactive materials.

Special instructions given to users:

1. Only assigned Dosimeter to be worn.
2. Not to tamper Dosimeter
3. Not to use the Dosimeter in washing machine for clothes washer.
4. To wear whole body badge at the chest level.
5. Dosimeter to be returned to the concerned RSO, if he is expected away on leave for a long period.
6. Not to expose Dosimeter purposely to radiation and moisture
7. Not to exchange Dosimeter with other employees.
8. Only trained persons will be allow to have Dosimeter
9. Radiation safety cell will offer whole body badge to be worn at the waist, underneath a lead apron, for pregnant women employees.
10. For evaluation TLD badges will be sent to Defense Lab once in 4 months.
11. Badges are to be returned in a specific time period and no delay is allowed.
12. Lost Dosimeters should be intimated immediately to Radiation Safety Cell.

The Radiation Protection survey is done during and after radioactive material handling to ensure that contamination is eliminated and necessary follow up is done. The record maintained for survey will have the signature of the surveyor, laboratory layout, areas surveyed, conditions of survey instruments, action to be taken and actions already taken for any untoward happenings.

All individuals routinely check their clothing during tea break, lunch break and after their day duty. The doors, cup boards, shelves, refrigerators, cell phones, switches, door mat etc are checked to prevent calamities.

Chemical fume hoods and special iodination cabinets are used for protection.

Caution Signs

Bold and clear sign boards are placed to caution one and all from radiological hazards. Each sign has the magenta, purple or black three bladed caution symbols on a yellow back ground with a suitable message like 'Caution Radiation area', 'Caution Radioactive material' etc.

All containers utilized for transporting or storing radioactive materials will have the label 'Caution Radioactive materials'. The type and quantities with date are also seen in all labels.

Empty containers are thoroughly made free from these labels.

If any equipment is age old or poor in quality, we experience air borne radio activity issues. The safety department will ensure that preventive measures are taken imminently.

Safe Transportation

Radiation Safety Cell checks while transportation to see that required quantity permitted by AERB is alone transported for storage.

Packages approved by AERB with radiation symbols are only transported.

The radiation level on the surface of the package is measured by RSO. The quantity of activity will be mentioned on the package. The logbook maintained by RSO will have details of storage, location, date, time of supply, name of the individual carrying the package etc for future reference and action. A copy of the details will be given to the Security Guard section at the gate.

Any error committed in disposal of radioactive waste will lead to severe health problems to outsiders. This will be questioned in the Parliament also.

Before procurement of radioactive material, they plan with vision for correct method for disposal as per rules designed by Radiation Safety Cell.

The waste storage area will be close to work area to avoid spillage problems when transferring waste to containers. Liquid wastes are kept in secondary containment. Each has a label on all sides cautioning radioactive material. Employees are cautioned not to place any radioactive material near the regular dust bins. The identification of containers will be clear and specific.

Without prior permission, different isotopes are not dumped in the same container. No radioactive waste is stored in the lab for a long period. Within a specified time it is disposed off.

The card board boxes used for disposal will be checked before throwing it into trash after disposal works.

Guard at the Gate

The frequent complaints received here is the danger due to spillage of radioactive material or solution. A trained lab worker will be engaged to clear it. Outside agencies are called for major contamination taking place in the premises. The names of persons engaged in the area where spillage occurred will be noted down. Immediate medical attention will be rendered to them after thorough medical checkup. No individual is allowed to go out from the spilled area until they are permitted by safety officials. No outsider will be allowed by the guard at the gate, when any untoward incident has happened inside that area.

All persons not involved in the hazardous area are sent out by the security guard. The room in which the contamination has happened is isolated or locked. A Radiation Safety Cell has been arranged to call them during any calamities. Without their permission no material should be thrown to dustbin or no one should leave without medical checkup.

Alarming Dangers

The anti-nuclear groups never fail to raise their voice about the alarming dangers to society through this nuclear technology.

It's true that nuclear storage in India needs up gradation.

****

Nuclear Security in India

Ministry of Home Affairs handles the threats in the outer region of nuclear storage yard with the team of State policemen. They also arrange armed men for transportation of nuclear wastes. They pass on information to Defense Ministry and AERB if any anonymous call of threat is received.

"Many reactors in India are unreliable and outburst of calamity will not be a surprise", say the environmentalists and social service doctors like Dr.Pughazhendi and leading Supreme Court Lawyer Dr.Buddhi Subbarao.

Villagers from Ghivali (19° 50' 44.5596" N, 72° 39' 52.6968" E) and Chinchani (19° 53' 10.5324" N, 72° 41' 20.4936" E) around Tarapur Atomic Power Station have agitated several times against the nuclear hazards, as it is the same model of Japan's Fukushima.

As the nuclear experts and scientists fear about these anti-nuclear activists and threats by terrorists, the vigilance on nuclear storage yards is always severe. They see that no untoward incident occur in these spots.

Security Posted

The security to protect the storage yard is done by State Police Force externally.

The CISF monitor the inner region with rifles and weapons.

The sophisticated biometric technique is used for entry of personals. Cement and steel barriers also prevent the entry of miscreants and terrorists. This arrangement with structural guards will avert any untoward incidents due to radioactive contamination. The entire store yard is managed by remote operated control room with various commands to prevent the entry of unauthorized elements. A team of technical expert staff check every week for any defects in this digital circuit network.

Safety Audit is done by AERB.

Many security guards are posted while transporting the radioactive materials and they do not disclose their road route to avoid the attack by the terrorists.

Weak Point

Before the appointment of a staff in this critical nuclear zone, his family background, criminal punishments given to him, if any, relatives living abroad and in neighboring enemy countries etc are scrutinized. These employees are checked for behavioral attitude with government divisions and general mental stability. Public and press have criticized that such caution was not taken in recruitment of contract laborers. This is a weak point in nuclear safety. There were instances when some temporarily appointed workers have committed grave crime in contaminating the atmosphere, which lead to poisoning of regular loyal employees.

Threats

The Fire service men will co-ordinate with the CISF men in time of emergencies. They are all trained in National Industrial Security Academy Institutions with all technical intricacies of radiological, chemical and nuclear dangers and threats from neighboring countries.

Ministry of Home Affairs handles the threats in the outer region of nuclear storage yard with the team of State policemen. They usually arrange armed men for transportation of nuclear wastes. They pass on information to Defense Ministry and AERB, if any anonymous call of threat is received. Suggestions given by them are also studied in depth by nuclear experts and CISF to improve the safety.

The Intelligence Bureau and State Intelligence experts analyze the possible nuclear hazards.

Disaster

National Disaster Management Force is posted in vulnerable areas all over India to protect nuclear storage threats. They also handle disasters due to Tsunami, flood and earth quakes. Usually it consumes more time to call them to nuclear storage area. Hence the State level security team should be trained and well equipped with modern weapons to safeguard the nuclear storage areas.

IAF men are not posted here, which is a weakness, even though they are more qualified to face terrorists.

Thermal Cameras and Fire Sensors used are to be modernized for better security.

Dangers

There is no proper record maintenance for the temporary staff engaged there, which leads to leak of information and nuclear attacks. The entry of men with smart phones is not checked and it may cause dangers.

Central Alarm Station, Watch Tower and Radiation Monitors are available. Scheduled check/inspection and non-scheduled surprise inspection are executed here by AERB to avert nuclear disasters. They check the gadgets, alertness of the CISF, road used for patrolling, Fire alarm systems, Cyber crimes errors in computer used in that complex, quality of weapons held by security and identification of suspicious elements amongst the contract laborers etc.

Still the nuclear security is not guaranteed in India!

****

Nuclear Emergency Plans in India

Emergency wing has the population data and demographic data to handle any nasty situation. The population data will cover floating population as well as the permanent residents around the nuclear plant. The list of regular employees and contract workers are given to them to handle the situation.

Even though the nuclear plants in India are erected, commissioned and operated as per the norms stipulated by the nuclear safety rules, accidents do occur under many critical conditions.

So, it is the primary aim of the nuclear scientists and operating team to manage the emergency and untoward happenings with care and caution. Or else the entire nation, environmentalists and opposition parties will pounce on the rulers.

Apart from the duties of the emergency units, the Nuclear Power Plant Chief Operating Officer has to chalk out emergency plans to save the public from calamity. There should be complete transparency in all their emergency activities and anticipated dangers in the plant. Significant guidelines are to be listed for the operating staff and nearby residents to escape from nuclear exposure.

The site details, plant layout, exit path, location of First-Aid room, emergency door details etc are to be displayed in all vital areas in and around the nuclear power station.

Brisk assessment of radiological condition, timely warning signals about radiation level to the nearby innocent people for safe evacuation and to get cover under shelter are done by the Emergency Action Team of the nuclear power plant.

Emergency wing has the population data and demographic data to handle any nasty situation. The population data will cover floating population as well as the permanent residents around the nuclear plant. The list of regular employees and contract workers are given to them to handle the situation.

The details of ground water usage for domestic, agriculture industries etc will be with them. With this they control the discharge rate for radioactive effluents into water and take preventive measures if it exceeds. Updated data on operating life of the plant and the changes expected in population and agriculture activities in due course is equipped by them.

The emergency action team inspects the houses nearby to check whether they are well built or thatched or mud structures. Emergency Planning Zone is first demarcated and all the public living in that zone should be protected if only they have amenities, strong shelter, stable windows and doors. Or else the Emergency Wing will be in panic situation to save the public from nuclear radiation. Schools, hospitals, orphanage, temples, marriage halls, tourist spots, restaurants, hotels etc inside the EPZ will be studied fully with plans to save from disasters.

Festivals will attract thousands of visitors and tourists from all over India and hence they have to be also protected. The Emergency Team will take into account the density of population and ways to tackle them.

The stock of medicines like Prophylaxis, Potassium iodide tablets etc is studied by this team and supply to the affected people with swiftness. In various stations there will be a team member with a nurse to distribute with instruction, when radioactive isotopes of iodine are suspected in the area around the plant.

During large gathering, the Primary Health Centre, Nuclear Calamity Clinic with sufficient doctors and assistants are to be posted by them in 3 or 4 locations. Cattle around the nuclear plant are taken care of with relevant statistics about it. The cattle milk when polluted will affect the users, whether young or old. Hence the emergency team member will arrange fodder and milk from the outskirts of EPZ. They also take care to see that a storage yard is arranged far away from the plant to face any drastic situation.

Transportation is another measure undertaken by the emergency units to prevent panic and unwanted accidents. All roads and railway facilities are studied in depth to evacuate the affected people. The condition of canals, bridges, boats, river routes around EPZ is studied by the teams, which manage the nuclear environmental hazards. Trucks, buses, autos, communication facilities, STD booths etc are planned before hand to face the situation.

Shelters, where the people in agony to be accommodated are preplanned by them with required toilets, drinking water facilities and food supply. The details about weather and wind are studied with Meteorological experts to obtain details to take protective actions when atmosphere is polluted with nuclear radiation. Detectors are installed to monitor the environmental condition.

The team is also equipped with Aircraft for Aerial survey to manage the nuclear contamination.

If only these plans are well executed, the public around nuclear power plants can be safe and sound.

Nuclear Emergency Units in India!

Even for a mild emergency in the site, the DAE and CMG are alerted by the nuclear plant engineers and administrators. Utmost care is seen that no leakage of information to the public or press take place here.

Department of Atomic Energy is the Nodal agency in India to manage man made radiological emergencies in the public domain.

For the past 28 years, the nuclear experts and scientists utilize the services of Crisis Management Group (CMG) to face calamities due to mal-operation or failure of nuclear power plant equipments.

This CMG will support and guide the local authorities and the National Crisis Management Committee (NCMC) in the calamity area. The executives and experts in Nuclear Plant operation and maintenance in Bhabha Atomic Research Centre, Nuclear Power Corporation of India Ltd, Heavy Water Board, Director of Purchase and Stores and AERB will be posted to govern the activities of CMG. Relevant data collection on hazards, communication and procurement responsibilities are supported by junior executives posted below these key experts in CMG.

Nuclear physicians play a vital role to advise the CMG to handle emergencies with drugs, treatment and hospitalization of patients. The intensity of radiation and preventive steps to be taken are suggested by scientists in BARC and Nuclear Power Plant Operation Managers.

For the area, which is 1.6km around the Nuclear Power Station, they earmark it as Exclusion Zone. In this area nobody will be permitted to move about. Strong fence will be there to prevent the entry of even cattle. 16km radius around the nuclear plant will be earmarked as Off Site Emergency Planning Zone.

The nuclear experts inside the plant handle emergencies like emergency standby, manpower emergency and nuclear plant emergency, which are happening within the plant only.

There may be emergency taking place within the Exclusion Zone and the public domain will not be affected by any radiological calamity.

If crisis is created to the public domain around the plant, then this Offsite Emergency is highly dangerous due to radiological release. Entire environment around the plant will collapse with death and injuries. So, at this instance the CMG, DAE, National Crisis Management Committee along with Local District Administrators like District Collector, Deputy Collector, Thasildar, Revenue Officer, District Medical Officer, Health Department and Police Department will break their heads to save the public.

Chief of NPCIL has to be well prepared for all the 3 types of emergencies with a team of experts.

Emergency Standby, Personnel Emergency and Plant Emergency are completely monitored by NPCIL and AERB. The plans are framed earlier and seen that no plan fails at the time of any disaster. To be on the safer side, the NPCIL Safety Board periodically conducts exercises and drills designed by Nuclear Safety Committee Team and AERB. The public and staff are trained to manage and escape from health hazards.

Even for a mild emergency in the site, the DAE and CMG Teams are alerted by the nuclear plant engineers and administrators.

The District Collector will take the responsibility to carry out Offside Emergency Plans with the support of NPCIL.

These plans will be chalked out well ahead of the commissioning of the nuclear plant. The Offsite Emergency Plans are executed once in a year and all the junior nuclear expert team in CMG, DAE, and Officials in Ministry of Health, Ministry of HRD, Ministry of Home Affairs and PM's office are communicated about the outcome. Copy of the drill details will be submitted to study in depth by key scientists and health experts for any modifications. Questions will have to be answered in the Parliament based on this report only.

Remarks by public representatives and environmentalists will be heard and alterations will be done immediately for the welfare of the society.

Mumbai and Delhi are vital locations to handle any emergencies because key officials related to Nuclear Power Safety are posted here. A fully equipped modernized Control Room is set up in these locations. VSAT, Email, FAX, Phone, Cell, Wireless and Hot line to PM's office arranged in these control stations are manned by the most reliable executives round the clock. Utmost care is seen that no leakage of information to the public or press take place here. Due to national security reasons, the army officials are usually posted here to communicate with care and smartness.

Totally 170 communication exercises and 110 emergency exercises are carried out, say the DAE officials. The District Administration Staff coordinate well with the support of Transport Department, Judicial Department, Communication Department, Meteorological Department and Revenue Department.

Emergency Control Room team will be alerted if any radiation leakage is observed in public domain, while transportation of nuclear wastes. This message will be imminently passed on to Crisis Management Group and DAE.

Local administrator, State Government Public Relations Department and State Publication department will prepare a booklet in local language to educate the people to protect from nuclear radiation. NDMA, NDRF and NDMA team handles all emergencies and relief activities. Based on National Disaster Management Act 2005 they are functioning to protect the society.

They have the NDRF posted in all States like A.P, M.P, U.P, Maharashtra, Assam, Bihar, Karnataka, Haryana, Kerala and Gujarat etc, where calamities are reported due to nuclear, manmade and natural disasters. There are 4 NDRF battalions and they have totally 5000 trained armed forces to tackle all situations. With the support and guidance of NDMA, they act to get relief from nuclear and radiological crisis. Just running here and there when nuclear emergencies occur is not a wise move.

Steps are to be taken for prevention of such calamities.

The responsibilities of the operating staff and people around the plant should be fixed well in advance at all levels.

Rehabilitation works are many after chaos and hence the NDRF team should be trained appropriately for that also. It is an expensive affair and the government should allocate huge funds for this cause.

Usually the disaster management focuses on training in natural made disaster relief works only and not on specific issues related to nuclear radiation.

So, special course is to be conducted to handle this aspect also.

New policy and rules are to be framed for individuals striving in these nuclear hazard areas with additional increments, out of turn promotions, bonus and special incentives, as they face unbearable injuries scientifically. They should be in the place where mock drills are conducted in nuclear power plants.

****

Grave Nuclear Crimes

Double checking of the manpower engaged in handling radioactive materials is needed in Indian nuclear power plants. The Central and State Police Force should join hands to nab the culprits. The machineries, which supply manpower should hold valid license, approved by AERB and State Police Force.

The appointment of staff on regular basis in nuclear power plant is usually done by giving advertisement in leading dailies, National Employment News Bulletins, Government office notice boards and NPCIL, DAE websites.

All the applicants apply on line after paying a stipulated fee by DD/ Cheque, Visa Card, Debit Card and Credit Card. After scrutinizing their applications, the applicants will undergo a test, interview, group discussion and health checkup to get selected.

But all these formalities are skipped for recruitment of contract laborers.

There are recruitment agencies to recommend them to get recruited after taking a huge amount as recruitment fees. With a big investment, these contract laborers enter nuclear power plants without any prior training or knowledge about nuclear hazards.

Usually cheap laborers from Bihar, Jharkhand, Assam, Nepal and backward areas in Andhra Pradesh are recruited. They are tough and mentally strong. They do not demand much pay but work just for the accommodation and food given to them by the nuclear plant authorities. Even for a shelter under a tent they will be satisfied and stay with family. Human Resource Department and Ministry for Contract Labor Act find difficult when cases are lodged by local police and environmentalists. These workers are paid less than the regular employee but the plant management will extract more work from these people and hence they have a grudge over the regular employees. Deccan Chronicle report says that in Kaiga Atomic Power Plant in Karnataka, a dissatisfied contract laborer disturbed the drinking water with heavy water and it resulted in health damages to 45 employees. This incident happened in 2009.

It was announced by the Atomic Energy Regulatory Board (AERB) in a press release that Tritium uptake of some workers at the Kaiga Generating Station (KGS) occurred on 24th November 2009. This was observed during the routine urine sample analysis of workers that is carried out regularly at all nuclear power plants that use heavy water. When checking two persons, they were found having Tritium in their body that can cause annual radiation exposure to exceed 30 mSv.

This calamity issue was raised by the opposition parties during question hour in the Parliament. The mode of selection of workers by NPCIL and DAE was opposed by environmentalists and anti-nuclear groups.

In MAPS, KARP and IGCAR the relationship between the management workers will be poor with many tussles and misunderstandings.

In 1997 the MAPS workers staged severe agitation when the management suspended 5 workers, who refused to work in high radiation areas. Their strike lasted for 25 days and hence the management appointed contract laborers to manage the show. These men had no knowledge about the nuclear hazards.

"Again in 2005 in IGCAR the workers stayed away from work spot as the roads to reach their homes were not good enough and at times of emergency they had to face peril. At that instance also the management appointed some unscrupulous lay men to handle the nuclear wastes and some operations inside the plant. These were the reasons for spreading of unknown diseases due to radiation till date around Kalpakkam", says Dr.Pughazhendi, an anti- nuclear agent in that region.

When unusual dreadful diseases was spreading near Rawatbhata reactor, the former Chairman of AERB, Mr.Gopalakrishnan reported that in 1970s and 1980s, cheap temporary laborers were engaged for cleaning operations of radioactive materials and there was no record about the number of young men and women engaged in it.

Again 6 contract laborers were engaged in digging a pit to check the burst lines below the earth in CIRUS and Dhruva reactors in BARC. They were working for 8hours for 2 days from Dec13th to 14th 1991, without proper protective equipments and they were pulled out and rushed back to their homes. They have been exposed to unsafe radiation levels, report the anti-nuclear activists.

In July 1991, white washing works had to be carried out in RAPS and a laborer engaged for it needed water for it. At that time barrels of heavy water kept near the wall of Heavy Water Upgrading Plant for up-gradation was noticed by him. Without knowing what it is, he used it for white washing the walls of the RAPS plant and he also washed his hands and face with it. This worker and the contractor, who engaged him, absconded after that dreadful incident. Entire team of nuclear engineers in that plant hurriedly took emergency measures to clean the walls. Since large areas in the plant are to be maintained with insufficient laborers, many works are remaining incomplete. Hence contract laborers were induced to do that job without any basic knowledge and the authorities had no time to teach them.

In BARC, where intense research is conducted in the field of nuclear field, some unidentified persons have intruded 25 times by-passing the security, say the environmentalists. But fortunately they could not approach the vital area of operations.

AERB has reported in 2010, that maintenance of nuclear and radiological materials was not handled properly due to inefficient untrained contract laborers in Mayapuri. Accounting of materials was not proper and no proper procedure was adopted for disposal, when its life was ended. There was communication gap between Delhi University and AERB regarding possession of radioactive materials.

The accounting of inventories should be in proper fashion in educational wings and hospitals, which use it for research and medical care purpose. If new comers and persons on contract basis are posted, they fail to follow the storing procedures for nuclear items and they don't record the date of receipt and disposal period. It should not be handled by all and only trained persons should be permitted to do it, say the environmentalists.

Double checking of the manpower engaged in handling radioactive materials is needed. The Central and State Police Force should join hands to nab the culprits. The machineries, which supply manpower should hold valid license, approved by AERB and State Police Force. Usually the regular staff will not be engaged in transporting, loading and unloading heavy materials. So, tough guys from local area will be seen. The State Police Team will have a complete record of photos, and crime activities of local vagabonds. Naturally these people may enter the nuclear plant for want of earnings or engaged by terrorists to spy.

The authorities in Hospitals, Fire services, Courier offices and Clubs near the nuclear plants should be in touch with security team to check the intrusion of unknown persons.

In case of theft or unusual incidents, the plant management should inform the local police and district administrator along with AERB. If they delay, then the culprits may escape from the scenario unnoticed by the local police team. Alertness is vital to overcome the crimes done by contract laborers.

Ghivali village is near TAPS and Poonam Hambire from this village laments that due to some carelessness and mischief, the people were affected by radiological hazards. As they were within 1.6km radius from TAPS, which has the same design as Fukushima, her nephew, faced death at the age of 8 due to lung cancer. No medical report could be obtained.

As his father was an employee, any protest will make his father lose his job in that deadly nuclear plant. Then the entire family will be in the streets. The mangroves around TAPS have dried up to cause loss to mango traders. The fish breeding has also been damaged due to heavy water mix with sea water, remark the fishermen.

Due to unscrupulous work culture in the nuclear plants, these damages are happening, say the villagers.

A contract laborer in TAPS has been exploited badly and he is suffering from joint pains. Along with him 29 contract laborers are suffering with no medical support from the plant management, reports the Village Chief.

"Prakash Ambhire aged 47 died due to eye cancer. He was a helper on contract basis and was not aware of nuclear hazards. He was an illiterate and no drills or exercise was rendered to him. No case study or medical report was prepared and entire matter is in dust bin, so that no environmentalist can take action against the plant. R.K.Gupta aged 73, who was an ex-employee in BARC Fuel Reprocessing Wing of Plutonium Plant, is undergoing various health hazards due to exposure to radiation. The contract laborers after the shift period are found dead along road side, uncared", says Ramakrishna Tandel, Secretary Maharashtra Machimar Kruti Samiti.

Contract laborers remain handicapped in a helpless way. Many employees have Psoriasis. The dosage for the employees should be lowered. At TAPS contract laborers are exposed to 1500 doses in a span of 2 months and employees 1000 doses in a year.

****

Safety Design for Kudankulam Nuclear Power Plant

The 2×1000 MWe Nuclear Power Plant is at Kudankulam along the shores of Gulf of Mannar near Kanyakumari, Radhapuram Taluk in Tirunelveli District in Tamilnadu State.

The Kudankulam Nuclear Power Project (KKNPP) consists of two units of 1000 MWe with two steam driven turbo generators supplied with turbine steam from two pressurized water VVER type reactors of Russian design. All design features is as per Russian and other International Standards.

NPCIL divided the entire project works into various packages and offered to known and experienced contractors.

With an initial investment of about Rs. 13,000 Crores the reputed MNC L&T was awarded M2, M4, M5, C5 and E1 packages directly by NPCIL. M3 and I1 were awarded to L&T by BHEL & ECIL.

Mechanical package consisted of erection of Nuclear Steam Supply System (NSSS) and Auxiliary Systems (equipment, piping & tubing works).

The M2 Package consisted of Seawater System (equipment & piping erection) and M4 Package covered Common Service (piping & equipment erection).

In the M5 Package, erection of Turbine and in M3 Package Civil works like Radioactive Waste Processing and Storage Buildings were done.

The vital erection of electrical equipment and systems were executed under E1 Package

Technical Details of Kudankulam Plant

Kudankulam Reactors are Water-cooled Water moderated Energy Reactor (VVER-1000) of Pressurized Water Reactor (PWR) technology, which is worldwide proven concept. VVER-1000 reactors are the most advanced reactors similar to the PWR's of western design. Nuclear fuel is charged into the Reactor, where the fission takes place and the heat is liberated. The liberated heat in the reactor is carried away by the light water from the reactor to steam generator by the main coolant pump. De Mineralized (DM) water acts as reactor coolant. The coolant water is used in the steam generators to produce steam. The steam is allowed to pass through turbine blades with high pressure and velocity. The turbine then rotates receiving the kinetic energy of the steam.

The generator, which is mechanically, coupled to the turbine also rotates and generates power. The back steam of the turbine is allowed to pass through the surface type condensers. The condenser will be having tubes through which sea water is allowed to flow and over which the steam flows. The sea water absorbs the heat from the steam and condenses the back steam. The condensate is recycled back to steam generator. The sea water after absorbing the heat from the back steam is discharged into sea and fresh water is drawn through inlet.

Larsen & Toubro was responsible for all major mechanical erection and installation works in this plant. This includes four mechanical packages and two electrical and instrumentation packages. The works done in M2 package included handling, transportation, pre-fabrication, welding, erection, inspection & testing of piping, instrument tubing and erection of associated equipment along with accessories for nuclear steam supply systems and nuclear auxiliary systems for Unit 1 & 2.

This main mechanical package involved heavy erection of critical nuclear equipment like Reactor Pressure Vessel, Steam Generators, and Reactor Cooling Pumps, associated piping in Reactor and Auxiliary

Buildings, various safety systems like Boron Injection System, Passive Heat Removal System and Containment Sprinkler System.

Erection and Alignment of NSSS Equipment

The reactor equipment (2 x 316 t each) works were erected in four stages. As it is a sensitive power plant and condemned by environmentalists, high degree tolerances limit (0.01mm) was maintained.

The main taboo they faced at the time of erection was that the clearance between the inner diameter of support structure and outer diameter of Reactor Pressure Vessel (RPV), which was only 10 mm and large volume of blue matching.

It should be noted that Steam Generator (8 x 307 t each) was erected in a confined space with the alignment in the range of 0.4 mm to 0.5 mm. Pressurizer (2 x 222 t each) works were also executed in confined walls with the tolerance of 8mm w.r.t containment wall and alignment in the range of 0.4 mm to 0.5 mm.

Above mentioned work was done using highly precision optical tooling survey instruments viz., Jig Transit. Main Coolant Piping (64 Joints of 990 mm OD), which is designed for a pressure of 176 kg/cm2 and temperature of 320°C has four quadrants.

It is linked to Reactor Pressure Vessel (RPV) assembly, Steam Generator (SG) and Reactor Coolant Pump (RCP). The piping is used to re-circulate the primary coolant thereby removing heat from reactor and transfer the same to steam generator. Pipes are made out of Low Alloy Steel with outside diameter of 990mm and 70mm wall thickness. It is special to mention that Austenitic Stainless Steel material is used for inside cladding. The total circuit had 32 weld joints per unit and the total weight of piping is approximately 260 t.

The pipes were supplied in pre-fabricated piping spool. To make it unique in India, a sophisticated welding method was followed for Main Coolant piping.

During the welding operation with hot temperature they used Controlled and Induction Heating technique, which was new to Indian erectors. Radiography of weld joint at 200° Centigrade was carried out with special fixtures and source protection measures.

The Works Involved

- Reactor Equipment - 13,125 t
  - Stationary Equipment - 3,036 t
  - Rotary Equipment - 652 t
  - Piping (SS/CS) - 3,100 t
  - SS/CS Valves - 1,180 t
  - Instrument Tubing - CS/SS - 1, 60,920m
  - Tube Welding - CS/SS - 9,000 Nos.

Turbine Erection

Handling of materials at site store/storage yard, transportation to site of work, erection testing and commissioning of turbine generator, feed pumps, piping and its auxiliaries were done under M3 Package.

Alignment of the turbine was done using Optical Aligning instrument and Fronius was used for orbital welding.

Sea Water System

Handling, transportation, pre-fabrication, welding, erection, inspection, testing and painting of sea water systems equipment and piping in Sea Water Pump house and Chlorination plant were done for M4 Package work.

EP fabrication and erection, anti corrosive painting, erection of gates and screens, GRP piping, vertical pump erection, Chlorination building works and tunnel piping were done with care in the pump house.

Coupon Hycote paints was used for the internal painting of sea water pipes. This is a new method adopted for the first time in India.

- Rotating equipment: 1,432 t
- Stationary equipment: 1,465 t
- CS piping: 2,108 t
- SS piping: 8 t
- Glass reinforcement piping: 35 t
- Titanium piping: 16.5 t
- Excavation and backfilling: 56,000 cu.m
- Structural steel: 500 t
- Cold insulation: 1,183 sqm
- Anti-corrosive painting: 55,683 sqmt
- Coating and wrapping: 4,000 sqmt
- Grit blasting and painting: 9,000 sqmt.

Common Service System

In M5 Package, handling, transportation, pre-fabrication, welding, erection, inspection, testing of indoor and outdoor piping and equipment of common service systems were covered. This consists of all indoor and outdoor piping works, which covered HDPE piping for portable water, fire-fighting system (SGA, SGC, SGE), buried piping (coating and wrapping included), boiler associated works, DM plant associated tanks and piping, and all nuclear ancillary building piping and equipment.

- Rotating equipment - 450 t
- Stationary equipment - 2,612 t
- CS piping - 2,452 t
- SS piping - 405 t
- CI piping - 65 t
- HDPE piping - 150 t
- AC piping - 255 t
- RCC piping - 900 t
- Excavation and backfilling - 238000 cu.m
- Coating and wrapping - 7,000 sqmt
- Supply, fabrication and erection of structural steel - 500 t

Salient features of Turbine

The 1000MWe turbine is a new model to India with steam condensing, four cylinder configuration (1 HPC + 3 LPC) with intermediate moisture separation and steam reheat, with rotational speed of 50 s-1 (3000 rpm), designed to drive directly an alternating current generator, mounted on a joint foundation with the turbine.

Excluding generator, the total length of turbine is found to be 40.65m. The turbine is provided with throttling steam admission. The steam from steam generator is supplied to 4 HP valve blocks. Each HP Valve block consists of one Stop Valve and one Governing Valve through which the steam enters the HP cylinder. HP Rotor is an Integrally Forged Rotor. LP rotors are also integrally forged rotors (blade length is 1200mm).

All rotors are provided with rigid coupling and supported by two journal bearings. The thrust bearing, mounted between HP cylinder and LPC 1 serves as a fixed point for the rotor.

• The steam from 4th stage of each LPC is extracted by 3 LP heaters 1.

• The steam from 3rd stage of LPC 2 & 3 is extracted by LP heater 2.

• The steam from 2nd stage of LPC 1 is extracted by LP heater 3.

• HPC turbine is of double flow cylinder having 5 stages in each flow.

• The steam is extracted from second, third and fourth stages of each flow for feed water heating by HP heater 5, 6 and de-aerator respectively.

• From HPC exhaust steam is extracted by LP heater 4.

• The steam from exhaust of HPC enters moisture separator re-heater.

• After separator re-heater the steam enters LP cylinder through LP valve blocks, which consist of 1 stop valve and 1 governing valve.

• LPC turbine is double flow cylinders having 5 stages in each flow.

• Mass of heaviest part of turbine is LP rotor - 80.5 t

- Mass of turbine - 1800 t

Important Features of Turbine

Quantum of works:
- -Turbine 4,028 t,
  - Generator 1,297 t,
  - Condenser 2,609 t
  - Feed pump drive Turbine & Aux 402 t
  - Feed pumps 169 t
  - Condensate Pumps 257 t
  - Cross over Pipes 283 t
  - Other Auxiliary Equipment 200 t
  - Other Piping 12 t
  - Insulation - 6,000 sq m
  - Supply and fixing of hilty bolt - 1,000 Nos.
  - Well erection - 824 Nos.

Safety

Safety system practiced at the project site was extremely stringent as it involved several heavy lifts, erection of critical equipment, hot processes and working in severe conditions like tunnel and confined space. L&T's team religiously followed the safety norms as per the requirement of the Atomic Energy Regulatory Board (AERB). L&T has been awarded the Project Director's rolling trophy for Best House Keeping by the client. All the pre-commissioning and commissioning activities were meticulously planned and implemented successfully. All the erection sequences and procedures were documented in 175 volumes. These different packages are progressing well, meeting the quality standards set by Russian authorities.

Apart from Mechanical packages – Instrument tubing works in reactor building was also in L&T's scope, which involved 160 km tubing works.

Electrical Scope E1 Package - Technical Aspects

- Main Power output system
    - Auxiliary power supply system
    - Classification of auxiliary power supplies
    - GIS
    - System grounding

Important Features of the Plant

- First 2 × 1000 MWe power plant to be erected in India
  with single transformer unit of capacity 417 MVA

  - First time in India having 4200m length of 400kV GIBD
  - Fully inside lined reactor containment, a first time specialty
  in India.

The tender specifications covered electrical system installation in all the buildings of KKNPP unit 1 & 2 such as 400kV GIS building, Switchyard Control Room building, Transformer Yards, Normal Operation Power Supply building, Common Station Power Supply building, Emergency Power Supply buildings and other Main Plant buildings such as Reactor building, Turbine building, Reactor Auxiliary building, Pump House etc., as well as supply of some electrical equipments, hardware, accessories required for the installation works.

System Description

The electrical system for the two units of Kudankulam nuclear power plant, each of 1000 MW, mainly consists of Power Output System and Auxiliary Power Supply System. The Power Output System is designed for evacuation of power generated at NPP. The Auxiliary Power System provides the power to NPP auxiliaries to carry out their assigned functions, in all operating modes of NPP.

Power Output System

The electrical power generated by the turbine generators at 24 kV, three phase and 50Hz is stepped up to 400 kV by the generator transformer and is evacuated through four 400 kV transmission lines. For reserve source of power to auxiliaries of NPP, Kudankulam NPP is connected to Tuticorin, SR Pudur and Kayathar through 230 kV lines. 400 kV and 230 kV lines are connected through Interconnecting Transformers.

400 kV and 230 kV Switchgears are of SF6 Gas Insulated considering the saline atmosphere at Kudankulam site.

400 kV is designed as one and half bus scheme. 230 kV is designed as two main bus schemes. Major components of the Power Output System for KKNPP include:

- 24 kV 1000MW Generator
- 24 kV 31.5KA Isolated Phase Bus duct (IPDB)
- 24 kV 28KA Generator Circuit Breaker (GCB)
- 3 X 417 MVA, 24/400kV Generator Transformer (GT)
- 400 kV 2000A Gas Insulated Bus duct (GIDB)
- 400 kV Gas Insulated Switchgear (GIS)
- 400 kV Transmission lines
- 315 MVA, 400 kV /230 kV Interconnecting Transformer (ICT)
- 230 kV Gas Insulated Switchgear (GIS)
- 230 kV Transmission Lines
- 1 phase 26.7 MVA Shunt Reactors (6nos)

Auxiliary Power Supply System

The main function of the auxiliary power system is to ensure the availability of sufficient power supply to the auxiliary system equipments during all modes of plant operation i.e. normal plant operation, shutting down the reactor, maintaining the reactor in safe shut-down state, containment isolation, reactor core cooling, preventing significant release of radioactive material to the environment and any other necessary functions.

For start up as well as for normal shutdown of both units of plant, station auxiliary power is drawn from the 400kV network through generator transformer and unit auxiliary transformers with the GCB open. During normal plant operation, station auxiliary power is drawn from the main generator from the tap-off to 24kV bus duct through the UATS. The reserve power supply is derived from 230kV grid through 230/6.3-6.3kV RATS and is used as backup power during non-availability of power supply from UATS.

The Common Station Auxiliary Power supply feeding the Common Station Auxiliary loads is derived from 230kV grid through 230kV/ 6.3-6.3kV CSATS and is backed up by the RATS. The system mainly consists of 6kV, 0.38kV and 220 V DC power supply sources to supply to unit auxiliary loads and common station auxiliary loads.

Auxiliary power supply system envisaged for KKNPP is categorized as per their functional requirements detailed below.

• Normal auxiliary power supply for each unit system including common station auxiliary supply system for the plant.

• Reliable auxiliary power supply of normal operation for each plant.

• Emergency auxiliary power supplies system for safety systems for each unit. 19 ECC

Electrical Works Installation, Description

- 24KV, 2800A SF6 Generator Circuit Breaker 2 Nos,
- 420/24 KV 417MVAGenerator Transformer 7 Nos,
- 24KV 31500A single phase Isolated Phase Bus Duct 1120 m,
- 315MVA 400/230/33kV Interconnecting Transformer 2 Nos,
- /31.5-31.5MVA, 24/6.3/6.3kV UAT/RAT/CSAT 9 Nos
- 27MVA 420kV Single phase Shunt reactors 6 Nos,
- 400kV 3150/2000A Single phase SF6 bus duct 4200 m 230kV,
- 2000A single phase SF6 bus duct 900 m
- 6kV 6.3MW Diesel Generator Sets 10 sets
- 6kV 3150A Isolated phase bus duct 2400 m
- 6kV 3150/1600A Switchgear Panel 522 Nos
- 6kV/0.4kV 1000KVA/400KVA Dry Type Aux. Transformer 107 Nos
- 0.38kV MCC Single Front Panel 444 Nos
- 0.38kV MCC Double Front Panel 668 Nos
- Cable Tray & Supports 1860 t
- Cables 7800 km
- HT & LT Termination 129000 Nos
- Light Fittings 21285 Nos
- Supply, erection, testing and commissioning of Line Protection Panels 7 Sets
- Lighting /Power Distribution Board 672 Nos
- Light Fittings 7331 Nos
- GI Conduits 344 km
- 650/1100V FRLS PVC Wire 800 km

Classification of Auxiliary Power Supplies Depending on the degree of reliability for requirement of safety, the power supplies are classified into three groups Group-1: loads that require uninterruptible power supply are kept in group-1 class.

These can be interruptible but not more than 20 seconds. These are supplied from 220 V DC, 380 V and 220 V AC power supplies normally derived from group-3 or group-2 power supply sources and whose back up source of power is from the battery banks.

The major components of Auxiliary Power Supply System are given as below.

- 63 MVA, 24/6.3-6.3kV UAT
- 63 MVA, 230/6.3-6.3kV RATS
- 230/6.3-6.3kV CSAT
- 6.3 MW, 6kV DG
- 6kV Switch Gear
- 1000 kva & 400kVa, 6/0.4kV and 6/0.433kV Auxiliary Transformers
- 0.38kV and 0.415 kV Switchgear/ MCCs
- Rectifiers • 220 V DC & 110 V DC Batteries
* Inverters
* 220 V DC distribution boards
* 220 V/380 V AC distribution boards
* 110 V DC distribution boards
* Cabling
* Cable penetrations

Group-2: AC supply to the loads, which can tolerate short interruption for a time defined by the conditions without effecting safety of the reactor, under all modes of the plant, including the condition of loss of supply from all off-site sources, is called the Group-2 loads.

These are supplied from 6kV, 380V and 220 V AC, 50Hz supplies, normally derived from group-3 power supply and whose back up source of power is from DG sets.

Group-3: Power supply to the plant auxiliary loads normally required under all modes of plant but which can tolerate prolonged interruption in the power supply, without affecting the safety of the plant, and do not require obligatory availability of power supply even after the actuation of the reactor trip system is called the group-3 power supply. These are fed from 6kV, 380V and 220 V AC, 50Hz supply derived from UATS, RATS or CSAT.

Principles of Control, Monitoring and Relay Protection System:

During operation, a system is understood as a set of devices of relay protection, control and monitoring of the electrical equipment, structurally and functionally involved.

At Kudankulam NPP, Independent System of Control, Monitoring and Relay Protection System are provided for many of the systems.

Gas Insulated Substation

Here switchgear apparatus for 220kV and 400kV equipment use the SF6 as Dielectric media due to many of its advantages.

- Though SF6 was developed in 1900, it was used as dielectric medium only after 1940
  - It is a powerful dielectric medium
  - Its dielectric strength is 2.35 times to that of air
  - It is 5 times heavier than air
  - It is non-toxic in pure form
  - Its dielectric strength is proportional to the pressure that it operates
- This is well suited for HV switchgears and GIS. The gas passing through the isolated bus duct acts as a powerful insulation media and this compact the size of the switchgear.

Grounding and Lightning Protection to achieve the requirements, a grid consisting of 70 sq.mm tinned stranded copper conductor has been laid by the civil contractors at a depth of not less than 0.5 m below the foundation, with a mesh size of 25m × 25m in the outdoor area and about 6m × 6m mesh size beneath the buildings.

GI - Earthling Conductors run to walls and floors inside the buildings, which form the main grounding lines, are connected to the buried copper conductors. All the metal components inside the building are connected to the grounding network. Lightning protection conductors in a mesh formation are provided over the building surface, along outer perimeter walls and on the roof and are connected to the grounding grid. Lightning arresters are installed on the stack top.

Instrumentation package involved executing the entire instrumentation work under the supervision of ECIL.

Scope of work included calibration and installation of 18,000 instruments such as Pressure, Flow and Level transmitters, RTD and Thermo couples, Pressure and Temperature gauges, Analytical instruments, Level Sensors, Seismic Sensors and Radiation Monitors,

Installation of 3000 instrument support stands, fabrication and erection of 200 t steel, 50 km of GI perforated tray for field routing and 176 km of cable laying and glanding work.

Civil Works C5 package involved 65000 cu.m of concreting works to construct De-Mineralized Plant (DM Plant), Boiler House, Solid radioactive waste building, Chiller units building, Compressor Building and Diesel Storage Pump House including NPCIL's Main Administration Building. 220 kV Gas Insulated Substation 21.

****

Nuclear Power Protestors!!

There are many protesting against the nuclear power technology in India.

Some are as follows:

Henry Diffen

After studying law in Madurai Law College, Henry Diffen served as lawyer for 28 years he is the executive Director of People's Caretaking Center, which fights for human rights issues.

He has strived for the welfare of the downtrodden poor people, who are cheated by industrialists by spoiling the environment and healthcare. He never hesitates to raise his voice against the nuclear tragedies.

He has international contacts and NGOs in India to overthrow usage of Nuclear power.

A.Muthukrishnan

Writer A.Muthukrishnan has written many articles on human rights under the title 'People at suicide point', 'Downtrodden', 'Environmental Pollution', 'Atomic devil', 'Nuclear equipments', 'Human rights' and 'Globalization evils'.

He has personally visited Kudankulam, Jaitapur, Tarapur and Kalpakkam to tell the people living around it that Nuclear is harmful for our livelihood.

He gathers various data about nuclear equipments and its dangers. He has also written articles on topics like 'Human beings in shit', 'A travel with friends', 'Tehelka Crime' and 'Don't hang Afzhal', which have been translated into various languages.

His anxiety to throw out Nuclear power from India is appreciated by all environmentalists.

P.Abdul Sahmed

P.Abdul Sahmed hails from Pudupattinam village near Kalpakkam.

He is the General Secretary for Humanitarian People's Party.

He has spread the nuclear dangers to the people around Kalpakkam.

He associates himself with TMMK and MMK parties to fight against atomic power installations, which destroys environment. He fights for healthcare, rehabilitation, education and job opportunities for the affected families around Kalpakkam.

Ponnuthai

Ponnuthai is a bold woman in the battle against Nuclear power.

She hails from VasudevaNallur in Tinneveli District and does Nature Farming with her 25 years old son. After she faced lots of miseries in marriage life she joined hands with Tamilnadu Women's wing and fought for women rights and economic development. As a good speaker she has been fighting against Kudankulam right from inception.

M.Vetrichelvan

M.Vetrichelvan, is an advocate from Chennai High Court.

He visits the villages around the Nuclear Power Stations to tell the people about the harmful radiations.

He participates in all meetings, group discussions and street discourses on nuclear dangers. He informs the people about the Nuclear Laws and its loopholes.

The science which supports the people in power and their relationship with foreign countries for globalization, fails to help the downtrodden and common man in the streets begging for food and water in India" says M.Vetrichelvan.

Safe Nuclear Plant

When so many activists are engaged in fighting against Nuclear power, an intelligent Tamilian V.Jagannathan is engaged in producing a safe Nuclear power plant.

He even submitted a paper on it in an International Conference for Nuclear Scientists held in Brussels in 2005 Aug 25th.

In that great conference Jagannathan and Usha Roy from BARC revealed the techniques of ATBR with a safe operation of Nuclear power.

As per their design, no dangerous radiation will take place when meltdown occurs as seen in Fugishima Atomic Power Plant.

Their invention revealed that dangerous wastes discharged will also be very minute. Usually in Nuclear Power plant Uranium or Plutonium will be used.

They have radiation dangers and for which all the environmentalists fear and fight against it. The Nuclear wastes from it will be hazardous to people for generations together.

But in the design presented by Jagannathan and Usha Roy, Uranium or Plutonium will not be used and instead they will use Thorium. This has 10,000 times lower radiation effect than Uranium.

"The hazardous wastes will also be thousand times less and also cannot act of its own. A catalyst is needed for it to act" says Jagannathan.

"Only with the support of Neutrons it can be operated. At times of Tsunami and earthquake, when neutrons are not used, then their activation will be stopped and hence prevent dangerous consequences", says Usha Roy.

It is considered as a clean isotope and can be immediately used as it is after excavation from soil.

In India Thorium is available in plenty. The scientists say that 12 million tons of Thorium is available in the world, out of which $2/3^{rd}$ is available in India. Along the sea shore it is obtained abundantly, say the scientists.

Even in Australia, Turkey, Norway and Brazil, Thorium is available. As lignite, coal and oil will be no more in due course; we had to depend upon Thorium for power production, say the energy experts.

"Even in Norway Statkraft has involved in producing power with Thorium, as it is considered as safe and environmental pollution free. It can be utilized even in the regions where it is thickly populated.

Norway has taken this decision too and many have supported Thorium for power production, even though they initially opposed all nuclear plants "say Jagannathan.

"One drawback which has made all countries to hesitate to use it is that the wastes discharged from Thorium based power plants cannot be used for Atomic Weapons, as we do, when we utilize Uranium and Plutonium. So as the wastes cannot be a useful one, they are reluctant to adopt it.

A certain groups of people have submitted a proposal to use Thorium in Kudankulam to alleviate the fear of the common man" says Usha Roy.

****

www.ingramcontent.com/pod-product-compliance
Lightning Source LLC
Chambersburg PA
CBHW021907170526
45157CB00005B/2001